And God shall wipe away all tears from their eyes; and there shall be no more death, neither sorrow nor crying, neither shall there be any more pain. **Revelation**

Without pain, without sacrifice, we would have nothing . . . This is your pain. This is your burning hand. It's right here. **Fight Club**

Pain pays the income of each precious thing. **William Shakespeare**

After great pain, a formal feeling comes – | The Nerves sit ceremonious, like Tombs. **Emily Dickinson**

But for pain words are lacking. There should be cries, cracks, fissures, whiteness passing over chintz covers, interference with the sense of time, of space; the sense also of extreme fixity in passing objects; and sounds very remote and then very close; flesh being gashed and blood spurting, a joint suddenly twisted – beneath all of which appears something very important, yet remote, to be just held in solitude. **Virginia Woolf**

Justice turns the scale | For those to whom through pain | At last comes wisdom's gain. **Aeschylus**

One pain is lessened by another's anguish. **William Shakespeare**

Nature has placed mankind under the governance of two sovereign masters, pain *and* pleasure. *It is for them alone to point out what we ought to do, as well as to determine what we shall do.* **Jeremy Bentham**

There is no coming to consciousness without pain. **Carl Jung**

I don't mind pain, so long as
Attributed to Oscar Wilde

Series 117

This is a Ladybird Expert book, one of a series of titles for an adult readership. Written by some of the leading lights and outstanding communicators in their fields and published by one of the most trusted and well-loved names in books, the Ladybird Expert series provides clear, accessible and authoritative introductions, informed by expert opinion, to key subjects drawn from science, history and culture.

MICHAEL JOSEPH

UK | USA | Canada | Ireland | Australia
India | New Zealand | South Africa

Michael Joseph is part of the Penguin Random House group of companies whose addresses can be found at global.penguinrandomhouse.com

First published 2020

001

Printed in Italy by L.E.G.O. S.p.A.

A CIP catalogue record for this book is available from the British Library

ISBN: 978–0–241–34553–5

www.greenpenguin.co.uk

Pain

Irene Tracey

with illustrations by
Stephen Player

Ladybird Books Ltd, London

What is pain?

Pain is our bodies' alarm system. The 'hurt' of pain is a perception produced by brain activity. It tells us that something is not right. Every animal experiences pain, including us – whether from a minor cut or bruise, having our hair pulled, sunburn, a hangover or perhaps aching joints. It warns us we're injured, or that something even worse might happen. It motivates us to act and stop the harm.

This is 'good' or short-term pain, and in a clinical setting it is called *acute pain*. It saves lives. We know this because some people cannot experience pain due to a rare genetic condition called *congenital insensitivity to pain*, or CIP. Historically, they died young, often after performing daredevil acts that produced horrific injuries, which of course they didn't feel until it was too late and the infection or bleeding caused death. A stark reminder of the importance of feeling acute pain.

Chronic pain is something entirely different. This is 'bad' or long-term pain and is defined as present or ongoing for approximately four months. It can be a symptom of some other disease or injury, or a disease in its own right. On average it lasts for roughly seven years, but sometimes it's more than twenty. Patients often have *comorbid* (additional) conditions, such as poor sleep, anxiety and depression. A staggering one in five adults have chronic pain, and it leads to enormous suffering. It is one of the largest medical health problems worldwide.

To understand pain involves delving into many different aspects of the brain and body, our consciousness and psychology. It is old in evolutionary terms and it is both a sensory and an emotional, subjective experience. This book will take us on a journey to understand better one of life's most ancient and important experiences: pain.

Early descriptions of pain

More than 2,000 years before we really knew that the brain was where thought, reason and perception (seeing, hearing, tasting, smelling, touching) emerged, Hippocrates (*c.*470–*c.*400 BC) said that the brain was the seat of private, 'subjective' things, such as joy, grief and pain. But it was René Descartes (1596–1650) who really put pain firmly in the brain. His famous illustration of a boy with his foot in a fire depicted a connection running, like a bell cord, from the foot to the brain, where it 'rang an alarm' to say, 'Hey, it hurts! Do something!' The problem with this representation is that it implies the signals from the site of injury to the brain are faithfully transmitted, unchanged. They are not. We now know that these signals can be turned up or down, like the volume on a TV, at various points along the journey from the injury to the brain.

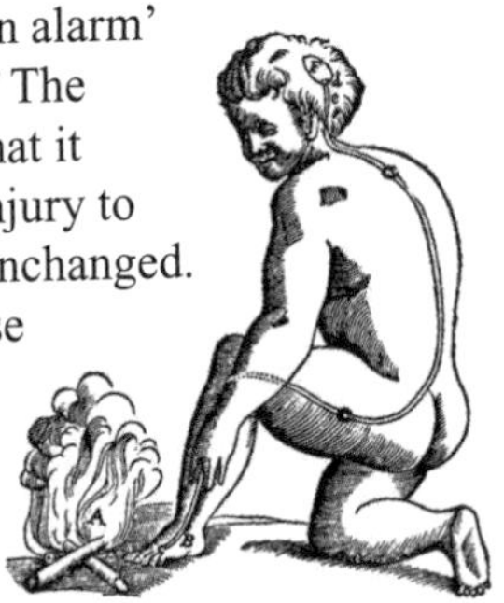

The Nobel Prize winning Charles Scott Sherrington (1857–1952) was an incredible neuroscientist who made profound discoveries related to pain. He revealed that specialized nerves just under the surface of the skin are activated only by things in the environment which hurt. He coined the term *nociceptors* to describe these 'pain-signalling' nerves. When triggered, they send signals to the spinal cord, which then relays them up to the brain, where 'pain' emerges as something you experience or 'feel'. Additionally, influences from the brain back to the spinal cord exist. These can strengthen or weaken the incoming *nociceptive* (pain-signalling) messages and form part of that volume-control machinery.

What hurts us?

Things that hurt us in the environment fall into three broad categories: the *mechanical*, such as a knife cut or a crushed toe; the *thermal*, i.e. burning heat or freezing cold; and the *chemical*, such as acid, chilli peppers, snake bites, nettle stings or the lactic 'burn' from exercise. These are examples of *noxious stimuli*: things that damage (or threaten damage to) normal tissues via a common end product – pain. But how? Well, those specialized nociceptors just under the skin and in the joints, muscles and other internal organs have special sensors called *receptors* or *ion channels*. Think of them as 'locks'. Particular components of a cut, a burn or a sting can 'unlock' these receptors or cause ions to flow through the channels. This activates the nociceptors, which signal to the spinal cord and then on to the brain, producing pain.

All species, from a simple sea slug through to humans, have the capacity to feel pain and react to it. Standing unexpectedly on a drawing pin, or touching a hot pan, produces a sudden jolt, perhaps an expletive, and a quickly withdrawn limb – Sherrington's reflex in action. These are *nocifensive* behaviours. Some animals, like that sea slug, might release a horrible smell or cloud of dye to deter predators. And spiders and snakes sting and bite, eliciting pain that warns predators not to mess with them again. These are clever *defensive* behaviours, using pain as a weapon.

The plant kingdom also produces chemicals that hurt – we think to prevent plants from being eaten. Though this doesn't stop us humans, who have an odd pain and pleasure relationship with foods like chillis, horseradish and mustard. Is the delightful smell of freshly cut grass a 'cry for help' to insects, as some think? A nice idea, but I'm not convinced.

Why does chilli taste hot and mint cold?

In the past two decades, the research teams of David Julius and Ardem Patapoutian, among others, have used molecular biology to discover a lot more about those *molecular sensors* needed to trigger impending pain. Many are *polymodal*, which basically means they can bind to multiple different noxious stimuli. Why is that interesting?

Well, let's take temperature as an example. We know it's possible to feel the difference between painful and tolerable cold, gentle and painful heat. Interestingly, the 'locks' that detect these different temperatures are from one large family, called *transient receptor potential* (or TRP) *channels*, and so quite similar. What is more remarkable is that several are polymodal and bind to something chemical in the world of plants, herbs and spices. We think of mint as 'cooling' precisely because the same nociceptor (TRPM8) is unlocked or activated by cold *therms* and by the chemical menthol in mint. The same goes for noxious heat: TRPV1 is activated both by hot therms and by the chemical capsaicin in chilli or jalapeño peppers. The poor brain cannot decipher which 'key' opened the 'lock' to trigger the nociceptive nerve, which is why we have the experience of chillis being 'hot' and mints as 'cool'! **Tip:** Capsaicin is a *vanilloid* chemical structure, which makes it soluble in fat. So drinking water with your curry spreads the capsaicin around the mouth, binding more nociceptors and making the pain worse. Eat something containing fat with it instead, like a raita or plain yoghurt, and the pain will be soothed.

There are many other types of receptors and ion channels outside the TRP family, which detect the noxious triggers produced by mechanical wounds or other chemicals. All ready and waiting to reveal injury or disease.

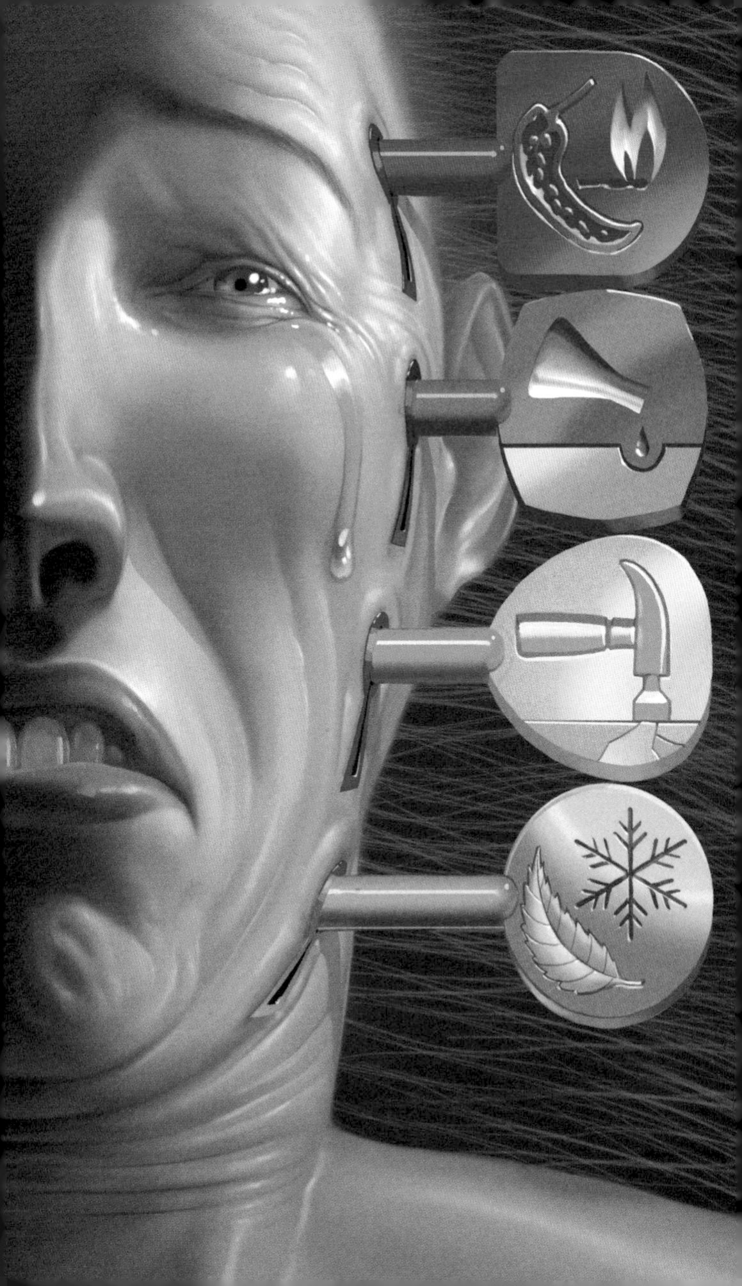

Fast pain, slow pain, no pain

All of our basic perceptions, like seeing or hearing, as well as pain, are encoded in a three-step neuronal system. The first step is *transduction*: the unlocking or triggering of the receptors or ion channels to activate the nociceptor. The second step is *transmission*, where the triggered signals travel along the *peripheral nerve* before entering the spinal cord. The third and final step occurs in the brain, which is *perception*: the 'ouch' of pain, and the moment we become aware that something hurts.

Peripheral nerves – i.e. those outside the spinal cord and brain, or central nervous system – comprise three main types: *Aβ code* for normal, non-painful touch; *Aδ code* for fast, almost 'sharp', pin-prick-type pain; and *c code* for slow, dull, throbbing and aching pain. Their job is to get the signals to the spinal cord, which they do at varying speeds: Aβ at around 50 metres/second; Aδ around 15 metres/second; c around 1 metre/second. Once there, nociceptive signals might be modified (like that TV volume) while connecting to other *nerve cells* or *neurons*, before they are carried on *nerve tracts* up to the brain.

Fast or 'first' pain is often precisely localized but fades quickly. It immediately attracts our attention, and often action. Slow or 'second' pain lasts longer, and reminds us something is wrong. It motivates behavioural responses to seek help and limit further injury. Some patients with congenital insensitivity to pain (CIP) have a genetic defect which means the transmission part (step two) doesn't work. This is why they don't feel pain, because the signals never reach the spinal cord and brain.

Recently, a new class of peripheral nerves, called *C-tactile nerves*, was discovered. They encode pleasant touch and activate during caressing. Research is determining if they also play a role in chronic pain.

START
100 SECONDS
FINISH

The gate-control theory

From Descartes's time, it was thought the spinal cord merely passed nociceptive signals to the brain, like passing the baton in a relay race. Then, in 1965, Ronald Melzack and Patrick Wall suggested something different. Observing that, when a child falls over and its mother or father gently rubs the injured area, there is magical relief, they proposed a theory to explain this apparent 'shutting of the gate' to the nociceptive signals coming from the site of injury.

The idea is beautifully simple. The normal-touch Aβ nerve fibres that are activated when the parent rubs the skin send 'touch' signals to the spinal cord and 'shut a gate' on the simultaneously incoming nociceptive signals from the c and Aδ nerve fibres. As these nociceptive signals are now blocked or inhibited by the 'gate', none are sent to the brain – and, hey presto, the child doesn't feel pain. This concept, which Melzack and Wall named the *gate-control theory*, was visionary as it allowed for the exciting possibility that *pain could be modulated* (turned up or down). The 'gate' in the spinal cord could control how much nociceptive signal is allowed up to the brain, in effect acting as a powerful 'gatekeeper' of the pain experience.

Surgeons exploit this idea with spinal cord stimulators for chronic pain patients. The stimulator does the job of the Aβ sensory nerve inputs (so there's no need to keep rubbing!) and forces the gate shut. Although it should be noted it doesn't work for all patients. The theory is also the basis of the TENS machine that some women use during childbirth. Here, electrical signals from the pads attached to the skin stimulate those Aβ sensory nerves to shut the 'gate' in the spinal cord on the labour pain signals. Again, it doesn't work for everyone.

The brain and pain

As Hippocrates said more than 2,400 years ago, pain arises in the brain. Without emergent brain activity there is no pain, just nociception. We need to be conscious to experience pain. Anaesthetized patients don't feel pain, so long as the anaesthetist does their job right.

Numerous tracts bring the signals from the spinal cord to a very old part of the brain at the cord's top, known as the *brainstem*, to arouse you (later we'll see it is key in modulating pain). Signals also go to a central 'relay station' in the brain – a bit like an airport hub – called the *thalamus*, and from there many other brain regions can be activated in a highly flexible way.

Think of a simple knife cut when chopping a vegetable. Ouch! – fast pain (those Aδ fibres). In an instant, almost without thinking, we're aroused, we locate the wound and recognize it as a cut (a mechanical injury) and as having a certain intensity or strength. The brain areas now active are our *sensory-discriminatory regions*. We have a sinking feeling; our attention is grabbed; we feel a bit stupid and frustrated, perhaps thinking we should chop more carefully next time. We are motivated to act, to run the cut under the tap, stop it bleeding and reach for a plaster. The brain areas doing all that are called the *cognitive-motivational-affective network*. While we finish dinner and do the washing-up, the cut continues to throb (those c fibres) as the quality and intensity of the hurt begins to evolve. All the time, the brain changes its activity pattern until the pain is gone in a day or two. In short, many brain regions activate in concert, yet in a highly variable manner, to produce this complex, multidimensional and changing experience.

Do babies feel pain?

During the 1890s, the German anatomist Paul Flechsig proposed that babies probably don't feel or remember pain. This was based on the observation that newborn babies' brains are not well developed (or *myelinated*, to be precise): also the folded outer layer, known as the *cortex*, is smooth. Until adolescence and early adulthood, it and other brain regions are still developing, myelinating, folding – being sculpted by our life's journey and our genes. So it made sense to Flechsig (and others) that perhaps babies' brains cannot process the nociceptive signals in the same way as those of adults. Ergo: babies don't feel pain.

Amazingly, this unlikely theory – or dare I say jaw-dropping myth – persisted. Fortunately, common sense, science, and modern imaging tools that allow us to see the brain in action have unequivocally overturned it. Babies' brains activate in a very similar way to adults' brains in response to things that the latter are able to say hurt.

In 1911, the renowned physicians Henry Head and Gordon Holmes proposed that the cortex was not important for experiencing pain. They derived their theory, in part, from observations that showed soldiers with cortical lesions still felt pain. Their hypothesis was soon overturned, though. We do need a cortex; even a baby's underdeveloped one allows them to feel pain and react to it with a yelp.

Plants have no brain or nervous system, which is why cut grass cannot feel pain. But what about robots? Can we train them to feel pain, or empathy for another's suffering, with a simplified nervous system – or at least react as if they have? Not yet . . .

Expressing pain: language, gestures, gender

There are fascinating historical, gender and cultural differences in how we choose to describe our pain. People often use metaphors drawn from their environment. For example, in India the heat of pain is likened to 'parched chickpeas', and the Sakhalin Ainu people of Japan describe 'woodpecker headaches'. Westerners, meanwhile, complain of 'gnawing', 'shooting' and 'crushing' pains.

Both nature (genes) and nurture (environment) play major roles in determining our pain threshold and tolerance. Some people really are more sensitive to pain, others less so; some people are stoical, others not. Sadly, pain has been highly gendered – negatively so towards women – but this is improving. Do men and women *feel* or *perceive* pain differently? Well, we're still working that out; it's hard to answer because pain is a subjective, personal experience. Certainly, more women suffer from chronic pain for reasons we're unsure of yet.

In the laboratory or clinic, we typically ask people to rate pain using two dimensions: intensity (how much) and unpleasantness (how bothersome). This complex interplay between the sensory and emotional components of pain means that no two experiences are the same, both within and between individuals. We commonly rate pain using numerical scales: for example, from an intensity of zero (no pain) to ten (excruciating). But is my 'ten' version of excruciating pain (after three childbirths) the same as your 'ten'? Our experiences make us feel and calibrate pain differently. Plus we're really bad at remembering. Can I accurately judge if my pain today is the same as yesterday's? This is the challenge of pain. It's malleable and hard to pin down.

%&#@!!
EEESH!
OH MY!
WAAAAH!!!!

'Measuring' pain – cultural and societal biases

There are three main ways we measure a person's pain intensity and unpleasantness:

1. Get them to talk about it using numerical or word-based scales, or questionnaires. For very young children, picture scales can be useful, e.g. sad to happy faces.
2. Observe their behaviour and make inferences about their pain and suffering from whether they grimace, guard an injured area, limp, etc.
3. If they are comatose, demented or under anaesthesia, then indirect *autonomic* responses (regulation systems we cannot control) are measured, such as heart or breathing rate changes and pupil dilation.

As pain is subjective and private, all these measures are imperfect. When we judge another person's pain, we also bring cultural and societal biases that are influenced by a rich tradition rooted in concepts like 'no pain, no gain', 'suffer in this world, reap rewards in the next', and so on. Although we have tremendous capacity for empathy towards another's suffering, let's be honest – sometimes it's hard not to doubt someone's rating or expression of pain. Whether it's assessing pain for disability claims in a law court, determining pain in anaesthetized, comatose or demented elderly patients, or measuring pain in premature babies or in animals, we must be mindful of such biases. The challenge of really knowing someone's pain should not be underestimated. New brain-imaging tools help us understand more objectively *why* someone's pain is a certain way. This is good for dispelling myths and negative biases or attitudes about pain.

Phantom limb pain

A classic example of pain that appears nonsensical, and so might not be believed due to bias, is *phantom limb pain*. This bleak and devastating chronic condition is usually felt by individuals who have lost all or part of a limb through accident or surgery, rather than by those born with a limb missing. Not only are they suffering from their missing body part, but they are reminded of its 'non-presence' by excruciating pain where it used to be, as well as other weird sensory experiences. Some people describe their symptoms as lightning bolts or electrical shocks, an icepick-like stabbing, or a burning fire. Imagine the distress.

An early theory proposed that the pain was caused when the brain, missing sensory inputs from the lost limb, amplified its signals, almost as if shouting out, 'Where are you?' An unlikely explanation. Another theory builds on observations that the signals in the brain are altered and rearranged as it adapts to the missing limb. It is suggested that these *plasticity shifts* (or *brain remapping*) cause the pain. Tricking the brain with a mirror box (or, these days, using virtual reality), so that it thinks the remaining limb is the missing one has been proposed as a way to normalize the brain patterns and remove the pain. Research is ongoing.

New theories are building on some of these observations and findings. In many instances, it is likely that nociceptive signals are still being sent from the damaged limb's nerves, where they were severed; and the brain, which does play a major part, mirrors these signals back – literally 'filling the gap' of the missing limb with pain signals. Blocking the damaged nerves with drugs can relieve the pain. But not always – so there's still more to discover . . .

Brain freeze: weird pains and pleasure

There are many weird and wonderful pains. *Referred pain* arises away from the site of injury. Phantom limb pain is a sort of referred pain. But it is more commonly known or experienced by people as a pain in the left arm during a heart attack; or, most familiarly, as 'brain freeze' – the headache when eating cold ice cream too quickly. During the body's development, as it is being 'wired up', the pain nerves coming from an internal organ like the heart muscle are mixed up with those from a particular section (or *dermatome*) of skin. When a heart attack occurs, the brain cannot tell which one sent the signals – so it always 'refers' the pain to the patch of skin, even though the signal is from the organ. The same goes for brain freeze: the nerves in the palate and roof of the mouth refer the pain to the forehead. The fancy name for this is *sphenopalatine ganglioneuralgia*. Luckily, the pain lasts only a few seconds – until the next gulp . . . So slow down!

Switching gears – can we make pain pleasant? Yes. Sports people, after an extensive training session or after an extreme event (e.g. running a marathon), reinterpret or reframe aching and sore muscles as 'good pain', because they associate it with something positive and rewarding. Even more extreme are sadomasochists, who actually find pain pleasurable. We don't yet know how, but in some way they change the value or meaning of the pain – reappraising or reframing it as something rewarding. The brain's ability to trick us like this is intriguing, and possibly offers new methods to alleviate pain in sufferers. Research is active in trying to understand how we can access and manipulate this powerful inbuilt brain system which can make pain pleasant.

21.04.62
Texting bleou fluor romo
blovc ehjo. Delln ne
JLANSF JUTDY KIVLOEST
TGK

Blocking pain: mind over matter

Einstein said: 'When you are courting a nice girl an hour seems like a second. When you sit on a red-hot cinder a second seems like an hour. That's relativity.' A clever quotation, which relates to this rule: *distraction alleviates pain*. Whether it is soldiers in battle, or sports people injured during the high arousal and distraction of fierce competition, they are often unaware of the hurt and pain – until afterwards. Then it hits. For chronic pain patients, sometimes listening to music or watching a gripping movie can provide some relief from their suffering.

This ability of distraction to minimize pain is actually very well understood. Remember Melzack and Wall's gate-control theory, which allowed for nociceptive signals to be modulated in the spinal cord? Another pair of pain scientists, Howard Fields and Allan Basbaum, built on this finding. What they described was a system in the brainstem that also connects to and modulates nociceptive signals in the spinal cord. It does it in two ways. One 'arm' is facilitatory and amplifies the signal, so more of it gets sent to the brain, making the pain worse. Whereas the other 'arm' is inhibitory and suppresses the signal, so less of it gets sent to the brain, reducing or totally removing the pain. The arms are like volume buttons, turning incoming signals from the injured area of the body up or down. Together with other brain regions controlling the brainstem, this system is called the *descending pain modulatory system*, or DPMS.

Going back to Albert Einstein: what distraction does is harness the inhibitory arm of the DPMS to turn the volume down. It accomplishes this, in part, by the release of *endogenous* (inbuilt) opioids, which help block the signals. Cool, eh?

Placebo analgesia: history

Placebo analgesia means getting pain relief from a pretend intervention or drug, such as a sugar pill. First, the patient has to be *conditioned* to *expect* pain relief after receiving the real thing; then they are 'tricked' with the sugar pill. The history of placebos is fascinating: the word dates back to the Middle Ages, where it referred to the Latin chants sung at one's funeral – often by hired 'mourning' monks! That is why, to this day, it is considered a derogatory term with connotations of deception and fakery. But this is far from the truth.

Based on his experiences treating soldiers in the Second World War and subsequent work, the physician Henry Beecher published in 1955 his discoveries on placebo analgesia (and other placebo effects). Without his patients knowing, he substituted saline syringes for morphine, and yet they still felt pain relief. Such observations led to the popularization of placebo-controlled randomized trials for developing new drugs: they have to perform 'over and above' the placebo effect to be approved.

Unfortunately, society and the medical profession then decided placebo responses were a bad thing. We went through a terrible period where 'placebo tests' were used to catch assumed liars – all based on a flawed understanding of the science and underpinning mechanisms (and with the usual social and cultural biases at play). But people weren't lying – placebos really work. Hippocrates and Galen, 2,000 years ago, understood the importance of the physician–patient interaction in influencing the latter's expectations and consequent pain outcomes. Harry Potter knew their power when he gave Ron Weasley a placebo dose of 'Felix Felicis' or liquid luck . . .

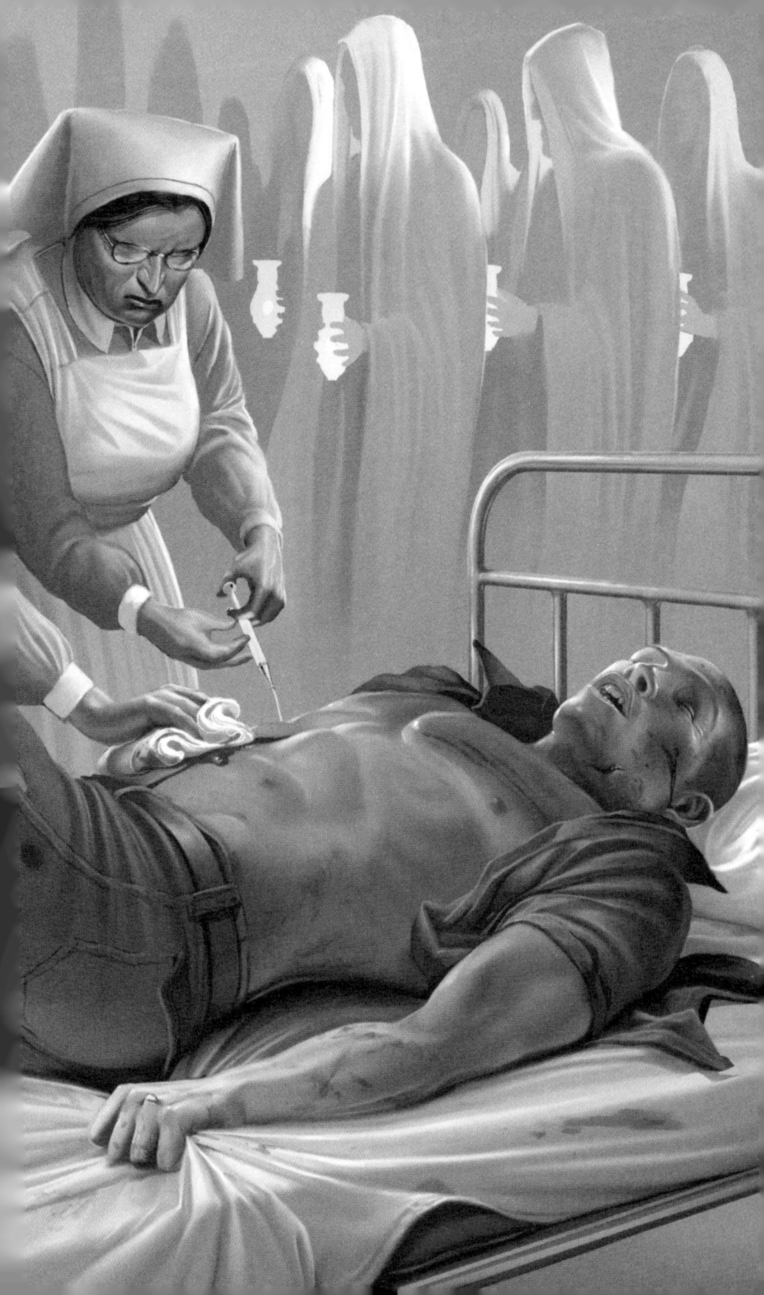

Placebos and nocebos: the science behind 'getting the pain you expect'

Many brain regions involved with attention, distraction and expectation are 'wired up' to control the inhibitory and facilitatory arms of the brainstem's descending pain modulatory system (DPMS). We know a lot about the neurochemistry here – *serotonin*, *noradrenaline* and endogenous opioids are all involved. In fact, if you give naloxone, which blocks the ability for opioids to work, you can prevent distraction (and placebo) from producing pain relief.

Placebo analgesia works by hijacking this system. Brain-imaging experiments have shown that, during placebo analgesia, areas in the front of the brain 'talk' to the brainstem – especially the *periacqueductal gray* – to drive the inhibitory arm that suppresses nociceptive signals in the spinal cord. As a consequence, it hurts less. Should we bring placebos back into modern medicine? Well, they require the physician to deceive the patient, so right now they're not appropriate, despite providing pain relief.

Relatedly, many surgical pain treatments don't get tested against a 'placebo surgery' – so it's not clear if the surgery itself has a benefit over and above a placebo effect. Recent studies, though, have proven some surgeries don't – i.e. they are just placebos. This is causing a real stir.

It's now time to introduce placebo's ugly other half: *nocebo*. A nocebo effect occurs when negative expectations or experiences, such as drug side-effects or thinking we're hurt when we're not, produce more pain. Believing a drug has stopped being given, even if it hasn't, makes the brain's anxiety 'amplifiers' override its analgesic effects to bring the pain back. You really do 'get the pain you expect'.

Chronic pain: facts and figures

Chronic pain is defined as pain that is present or ongoing for approximately four months. It produces untold suffering in one fifth of all adults, and is more prevalent in the elderly and in women. It is a huge financial burden to society, costing roughly $600 billion in the USA and €200 billion in Europe each year for its treatment and management and for the loss to the economy from sufferers unable to work.

Chronic pain can be a symptom or a disease. There are three main categories: *nociceptive* or *inflammatory*, e.g. osteoarthritis and rheumatoid arthritis; *neuropathic* (nerve injury), e.g. diabetic painful neuropathy, multiple sclerosis, stroke or traumatic nerve damage; and *idiopathic* (cause unknown) or *functional*, e.g. fibromyalgia or irritable bowel syndrome. Most sufferers have other comorbidities, such as anxiety, depression, catastrophizing (exaggerated negative thinking), insomnia, fear of moving, and *anhedonia* (inability to enjoy everyday things). Rarely, some chronic pains are caused by a 'gain of function' mutation, like *erythromelalgia*, which results in horrible constantly burning legs and feet. (An opposite 'loss of function' mutation in the same gene causes that congenital insensitivity to pain we spoke about earlier.)

Sadly, many of our drugs are not very effective. Patients can become overly dependent on and even addicted to opioids. After a while, opioids produce mostly side-effects without necessarily any pain relief. Currently, the USA and some other countries have an opioid epidemic, which is killing many every day. We need better ways to alleviate chronic pain. Unfortunately, there is no simple genetic or lifestyle explanation yet for why some people develop chronic pain.

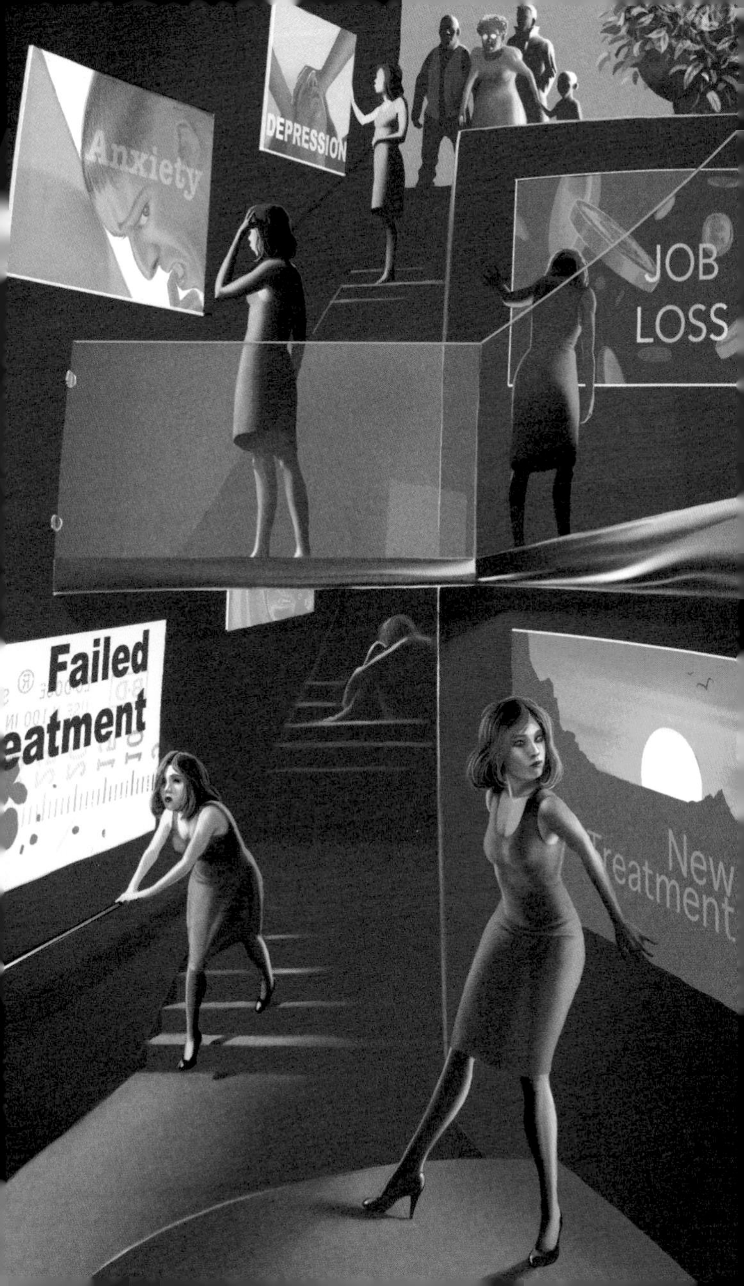

Anxiety
DEPRESSION
JOB LOSS
Failed
eatment
New
Treatment

Chronic pain: symptoms and a bit of science

Patients with chronic pain – and acute pain sufferers, too – can experience a range of unusual symptoms that are hard to describe. Common ones that occur spontaneously, or are triggered by touch or movement, include burning, electrical shocks, pins and needles, freezing cold, and sensations of creepy-crawlies walking over skin. And then for many there's the ever-present, relentless throbbing or aching pain. Chronic pain patients can also suffer from abnormal sensitivity to stimuli. This can take the form of *allodynia*, where the normal touch of clothes or bedsheets on uninjured skin causes pain, and *primary* or *secondary hyperalgesia*, where the perception and reaction to something painful is amplified.

Have you ever been sunburnt? What was once a lovely warm shower is now unbearable, or that lovely evening outfit touching your skin becomes agony, and you regret staying on the beach that extra hour. Or perhaps a skin graze is tender and painful for a while? These are examples of the skin becoming *sensitized* (a fancy term for 'more tender') after injury. There is a complex release of substances in response to tissue damage, and this so-called *inflammatory soup* activates and lowers the firing of those nociceptors, making them more easily triggered by things that are not normally painful, like the touch of clothes or warm water. Once the inflammatory soup settles down (naturally, or sped up by drugs), the firing threshold rises and that shower is pleasant again. This is acute and NOT chronic pain. But understanding it, and the longer-term role the immune response during inflammation has on pain, will help us to apprehend some of chronic pain's features.

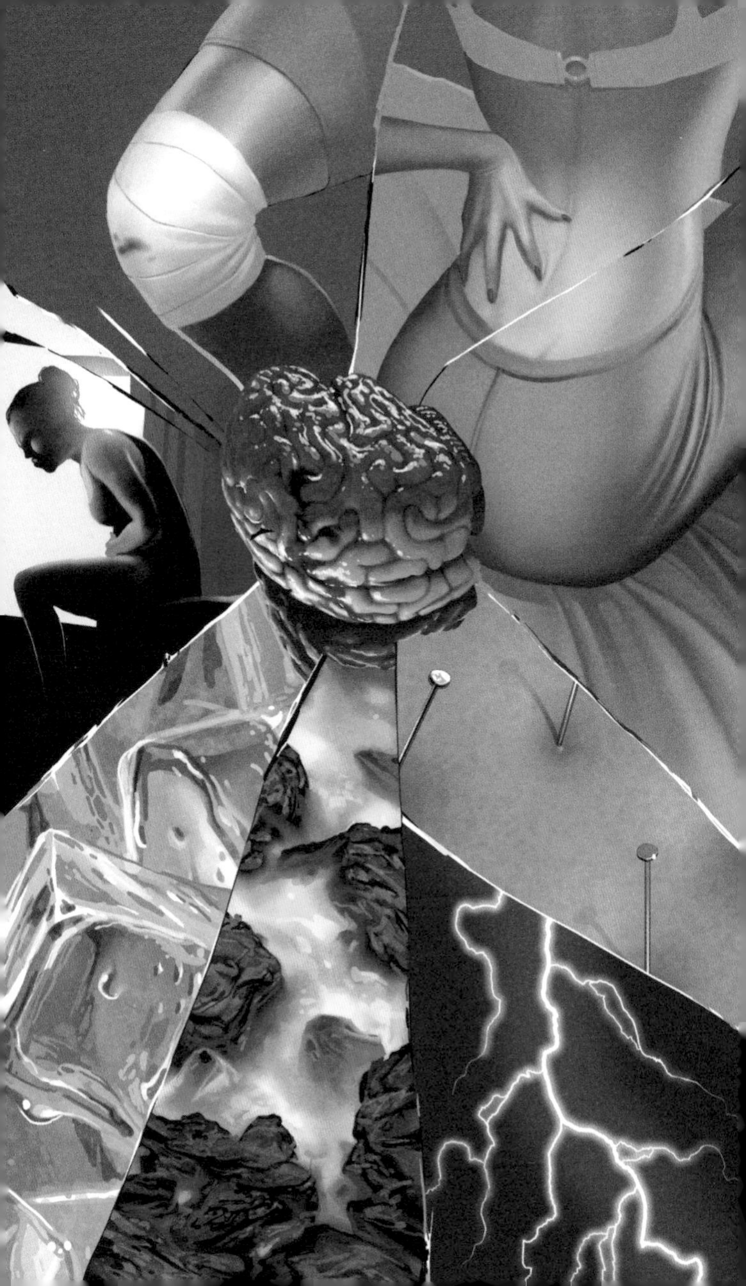

Chronic pain: a bit more science . . .

In the transition from acute to chronic pain, several 'mechanistic' changes contribute to the latter's development, maintenance and exacerbation. They occur at the genetic, cellular, nerve-wiring, brain-network and psychological levels. This means that all parts of that three-step neuronal system we spoke about earlier play a role in the chronic pain condition.

Unlike other nerves, which generally stop working when damaged, injured Aδ and c peripheral nerves switch on and start firing – constantly. The signals they send to the brain saying, 'I'm hurt', are a major problem in chronic pain. We need to stop them. Recent work has shown that *nerve growth factor*, or NGF, is released in the immune response after injury and is increased in several pain states. NGF causes *peripheral sensitization* and is a potentially exciting and novel target for treating pain.

Another really important mechanism underpins several key chronic pain symptoms and makes them worse. Discovered by another pain scientist, Clifford Woolf, it is called *central sensitization*. This is a fancy term to describe amplification by the spinal cord (as part of the central nervous system) of incoming signals, further heightening the pain. When combined with the lacklustre inhibition or overzealous facilitation of a dysfunctional descending pain modulatory system (DPMS), central sensitization is a key mechanism in chronic pain and contributes, in part, to allodynia, hyperalgesia and increased ongoing pain.

And if that wasn't bad enough, then focusing on one's pain, as well as comorbid depression and anxiety, also contribute to it worsening at the brain level. The challenge is to stop damaged nociceptive fibres from firing and to calm down all these amplification systems.

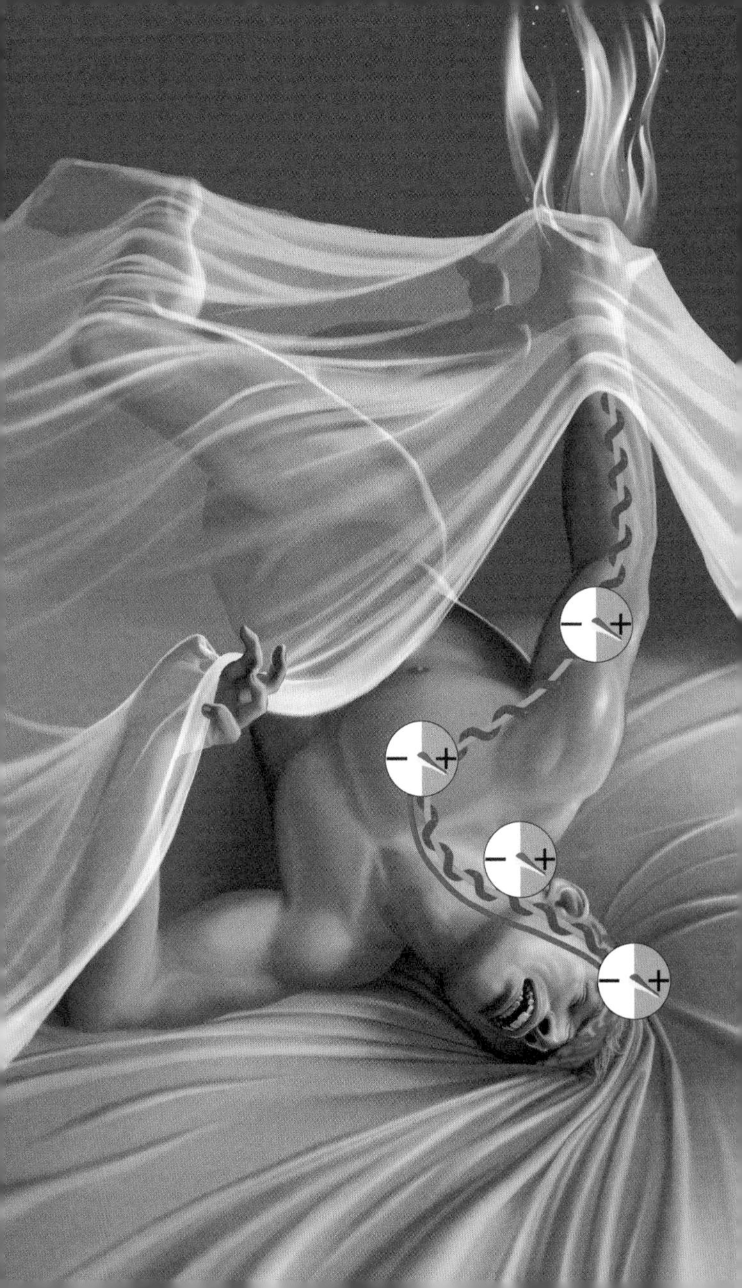

Headaches: hangovers and migraines

Let's be honest, hangover headaches are not uncommon. Interestingly, they have some symptoms in common with migraines: sensitivity to light and sound, vomiting and dizziness and fatigue, alongside a throbbing headache. The origin of hangovers and migraines and the explanations for headache pain are, however, quite different. In contrast to hangovers where alcohol and dehydration underlie headache, migraines can have a genetic basis. Migraine pain is debilitating, recurrent, can last several days and in some patients is preceded by visual disturbances.

Where do the nociceptive signals that produce hangover and migraine headache pain come from? Well, the various 'skin-like' protective coverings of the brain (the *meninges*), the blood vessels that supply it, as well as the skin of the face, neck and head, have a lot of Aδ and c nociceptors within them on pain fibres arising from the large *trigeminal nerve*. One thought is that alcohol, and possibly the *congeners* in red wine and dark liquors, cause vasodilation (swelling) of blood vessels and this activates nociceptors in the meninges to signal pain via brain activity in pain networks. Congeners, alcohol and importantly the physiological consequences of dehydration might also directly activate the trigeminal nerve to signal pain. To be honest, it's still not clear.

For a long time, this *vascular hypothesis* was also believed to explain migraines. But we now think it isn't the full story. Attacks may originate in the brain, trigger the trigeminal nervous system and sensitize pain-generating networks during migraines. We now know that substances, such as *calcitonin gene-related peptide* (CGRP), which are released by the trigeminal nerve when it is activated are involved in migraine pain. New treatments that neutralize or block the effects of CGRP are proving effective for some sufferers.

Emotions as 'amplifiers' and psychogenic pain

Ever gone to the dentist and been terrified by the expectation of pain from the drill? Inadvertently, you'll have made the pain worse. The same happens if you're sad or depressed. Research has shown that feelings and emotions like anxiety, fear, sadness and depression act as brain-based 'amplifiers' of pain. Not only is the volume of incoming signals increased, but the precise combination of brain regions activated is altered as well, powerfully influencing the pain experienced. This accounts, in part, for the 'mismatch' or discrepancy often observed between the extent of tissue damage visible and what the sufferer actually reports feeling. These 'hidden' emotional amplifiers are as important as the injury itself in determining pain.

A person can even experience pain without any nociceptive input or tissue damage at all – *psychogenic pain*. Relatedly, have you ever felt empathy for another person's pain or suffering? Felt hurt at being socially excluded? Been dumped by a partner? Or grieved the loss of a loved one? In these extreme and catastrophic situations, we turn to words like 'pain' to describe our suffering. Why? Partly because it really is pain-like. Studies show that feelings of empathy, social hurt and vicarious pain produce activity in some of the same brain regions as nociceptive-driven pain (although there are some important differences, too).

As a society, we are perhaps more comfortable with a nociceptive-driven model of pain – we prefer to see the origin of pain in blood, cuts and gore. But this type of pain is not more important than emotional or psychogenic pain. Pain is pain, no matter the source. Its very definition allows for multiple origins.

Treatments for acute and chronic pain

We have a large armamentarium for treating pain. But although acute pain is reasonably well managed, chronic pain is not. Around 60 per cent of chronic pain patients get virtually no relief from any of the treatments currently available, and the relief gained by the other 40 per cent isn't adequate or full. This is a bleak situation and we desperately need new and better therapies.

Pain is treated in four main ways:

1. *Pharmacological* remedies, i.e. drugs.
2. *Psychological* and talking therapies: e.g. *cognitive behavioural therapy*, *mindfulness* and *acceptance therapy* – each targeting different aspects of being in and living with chronic pain.
3. *Physical* therapy: e.g. rehabilitation and exercise to remove the fear of movement.
4. *Surgery*: e.g. replacing an arthritic joint, or inserting a stimulator – either of the spinal cord (to exploit its 'gate' controls) or of various brain regions, such as the periacqueductal gray (to drive the inhibitory arm) or the anterior cingulate cortex (supposedly to stop the patient minding about the pain).

Generally, a combination of these treatments is recommended, as they each target different aspects and mechanisms underpinning chronic pain conditions.

Complementary alternative medicine (CAM) can also help. With *hypnosis*, the science of how it works is still unclear. *Acupuncture* is an intriguing one. It has been used successfully in Eastern medicine – and surgery – for at least 2,000 years. We still don't really know how it works either – research continues.

HYPNOSIS
ACUPUNCTURE
DRUGS
OC
30
10
15
20
25
30
UNITS
THERAPY
SURGERY

Nature's pharmacopoeia

The two oldest pain drugs, used since antiquity, are aspirin, or *acetylsalicylic acid* (whose precursor, *salicin*, is found in willow trees), and morphine (from the seed capsules of opium poppies). These days, aspirin is largely replaced by ibuprofen (if there is inflammation) or paracetamol (if no inflammation). The biology behind *nonsteroidal anti-inflammatory drugs* (NSAIDs) like aspirin or ibuprofen is simple. They block key enzymes (*cyclooxygenases*, or COX) that form *prostaglandins*, which are vital components of the inflammatory soup that activates and sensitizes pain pathways. Because paracetamol inhibits COX only in the central nervous system, it has little effect on inflammation. It can be used to reduce fevers and mild to moderate pain – but how it works is still a bit of a mystery.

Society's love–hate affair with the potent drug morphine – named after Morpheus, the Roman god of dreams – began when the German chemist Friedrich Sertürner first extracted it from opium (or 'poppy tears') in the early nineteenth century. For acute pain, morphine and its variants are still widely used and efficacious. With chronic pain, though, tolerance, habituation, addiction and variable efficacy mean the use of morphine and other opioids is highly problematic.

Amazingly, we make our own 'morphine': endogenous opioids, released during placebo analgesia, or as *endorphins* that produce a runner's 'high'. We even make our own cannabinoids (*endocannabinoids*); but there is no good evidence yet that any of the cannabinoids in the cannabis plant can alleviate chronic pain, despite cannabidiol oil and medical marijuana being commercially available. Some snake venom proteins (*mambalgins*) and Botox (from the bacterium *Clostridium botulinum*) are more recent discoveries with potential pain-relieving properties.

Analgesia, anaesthesia and pain

Other important mainstream analgesics include antidepressants and anticonvulsants. The latter dampen down neuronal excitability, and some antidepressants reset the neurochemical imbalance that often occurs in the brain's modulatory systems (e.g. the DPMS). Lately, *biologics* that block nerve growth factor (anti-NGF antibodies) have provided further options.

Arguably, the best way to remove the conscious experience of pain is to remove consciousness itself. This is best restricted to controlled and reversible situations such as surgery. Today, millions of general anaesthetics are given annually to achieve the *triad of anaesthesia*: no pain, no awareness, no movement. Despite scare stories of patients being aware, feeling excruciating pain and not being able to move or speak when the surgeon's knife is wielded, this traumatic experience is, fortunately, extremely rare.

Oddly, despite the first general anaesthetic being used in 1846, we still don't know precisely how they work. Many different chemicals and drugs produce anaesthesia (unconsciousness) – so there is no single cause or simple mechanism. Some anaesthetics are given intravenously (e.g. propofol), others via inhalation (e.g. halothane). Using advanced brain-imaging tools during anaesthesia, we can 'watch' brain networks shutting down – like lights being turned off – to learn how unconsciousness emerges and pain diminishes.

Unlike general anaesthetics, local anaesthetics stop pain only in the regions where they are administered. This involves temporarily blocking ion (sodium) channels in nerves, which stops the transmission of pain signals into the central nervous system. Local anaesthetics are safe and effective, and widely used in routine medical and dental practice.

Cl
Br
F
O
O
F
F
F
F
F
F

The dark, bright and future faces of pain

Unfortunately, pain is also exploited as a tool of abuse. Torture is employed as a weapon of war or interrogation, and pepper spray is used for riot control by some law enforcement agencies. Some people even put capsaicin on horses to make them allodynic (painfully sensitive), so they lift up their legs more promptly in competition.

On the bright side, scientific research is discovering new and exciting ways not only to understand pain but also to treat it. For example, the ion (sodium) channel that doesn't work in people with congenital insensitivity to pain (CIP) is now being targeted to stop the transmission of pain signals in non-CIP patients. Nerve growth factor (NGF) and calcitonin gene-related peptide (CGRP) are being neutralized by new biologics which can treat inflammatory pain and reverse or prevent migraines. And the ability to block those newly discovered transduction sensors, which mediate such sensations as pressure, temperature and taste (like the TRP family), has the potential to relieve pain – providing hopes of a new class of safe and effective medications.

There is also an increasing use of digital health in pain management. Brain–computer interfaces and neural stimulators that target particular brain or spinal cord networks involved in the sensation and appraisal of pain stimuli are now becoming options for patients with chronic pain conditions. The final frontier is understanding what makes someone vulnerable to developing chronic pain, or resilient against it – as well as predicting what treatment is likely to work best for each sufferer.

Science is busting pain myths, changing society's attitudes towards pain and giving us new treatments to treat chronic pain. Good times ahead.

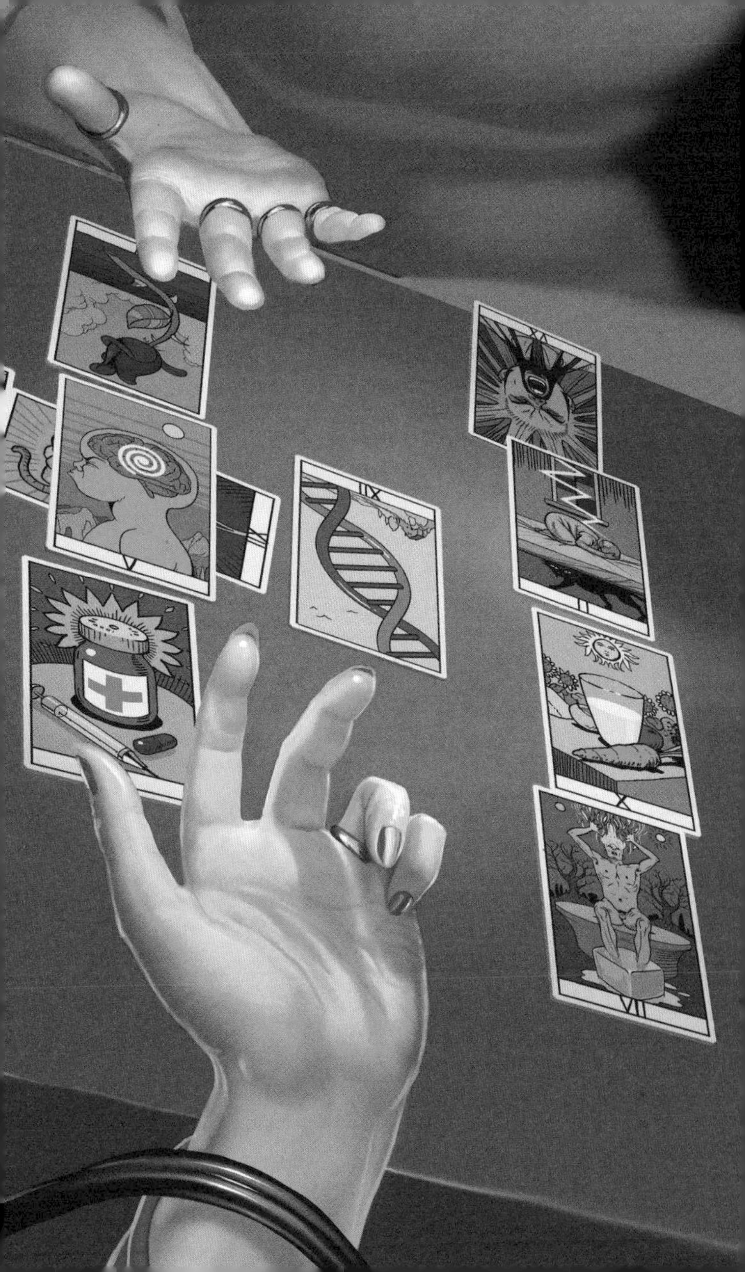

Further information

GENERAL

Joanna Bourke. *The Story of Pain: From Prayer to Painkillers*. Oxford University Press, 2014

Walter A. Brown. *The Placebo Effect in Clinical Practice*. Oxford University Press, 2013

Ronald Melzack and Patrick D. Wall. *The Challenge of Pain*. 2nd edn, Penguin, 1996

Pain Exhibit online art galleries. painexhibit.org/en/

Elaine Scarry. *The Body in Pain: The Making and Unmaking of the World*. Oxford University Press, 1985

Susan Sontag. *Regarding the Pain of Others*. Hamish Hamilton, 2003

Irene Tracey. *The Anatomy of Pain*. BBC World Service. January–February 2018 (www.bbc.co.uk/programmes/w3cswdkg)

Irene Tracey. *From Agony to Analgesia*. BBC Radio 4. August 2017 (www.bbc.co.uk/programmes/b0925604)

Irene Tracey. How Pain Works. *BBC Science Focus*. July 2017

Nicola Twilly. The Neuroscience of Pain. *The New Yorker*. 2 July 2018 (www.newyorker.com/magazine/2018/07/02/the-neuroscience-of-pain)

Patrick Wall. *Pain: The Science of Suffering*. Columbia University Press, 2000

ADVANCED

Pankaj Baral *et al*. Pain and Immunity: Implications for Host Defence. *Nature Reviews Immunology*. 2019; 19(7): 433–47

David J. Beard *et al*. Considerations and Methods for Placebo Controls in Surgical Trials (ASPIRE Guidelines). *Lancet*. 2020; 395(10226): 828–38

David L. Bennett *et al*. The Role of Voltage-Gated Sodium Channels in Pain Signaling. *Physiological Reviews*. 2019; 99(2): 1079–1151